The Poor Man's Ray Gun

(Deadly Rays)

by
David Gunn

Desert Publications
El Dorado, AR 71731-1751 U. S. A.

The
Poor Man's
Ray Gun

(Deadly Rays)

© 1996 by David Gunn

Published by Desert Publications
P.O. Box 1751
El Dorado, AR 71731-1751
501-862-2077

ISBN 0-87947-155-7
10 9 8 7 6 5 4 3 2 1
Printed in U. S. A.

Desert Publications is a division of
The DELTA GROUP, Ltd.
Direct all inquiries & orders to the above address.

Foreword

When the founding fathers of our great country drafted the constitution they included many checks and balances of power, so that one branch or department of government could not go unchecked. The President, Congress, The Supreme Court, and the States all oversee one another and limit each other's power so that no one department or entity becomes too powerful.

If we could travel back in time to the drafting of the constitution, we would find that the people involved had just kicked King George and his tyrannical government out of America. They were beaten by citizens, oppressed, took up arms with the British Empire.

Now when you view the first and second amendments through the eyes of the writers, you can see that they also are checks and balances.

The first amendment, freedom of speech, is important so that anyone who wants to speak out against government, may do so without fear of repercussions or persecution.

The second amendment, the right to bear arms, was not written to give hunters the right to hunt, as so many politicians would have us believe. The second amendment is the ultimate check and balance of our constitution. In case the government they just created became as oppressive as the British empire had been to them.

I am not saying we should take up arms against our government. I just don't think we should lay down our weapons and limit our Constitution with gun control legislation.

So long as authors such as myself, and publishers like Desert Publications exist, no man shall go unarmed.

Warning

This book is for entertainment purposes only. Building a maser would be a violation of many state and federal laws, as well you could kill yourself!

Contents

Chapter 1

Masers

Maser is an acronym that stands for Microwave Amplified Stimulated Emission of Radiation. As the name
suggests, it is very similar to a laser which is an acronym,
for Light Amplified Stimulated Emission of Radiation.
There are three different types of radiation: mechanical
radiation such as sound waves, electromagnetic radiation
as in radio waves, and particle radiation where a radioac-
tive material like uranium emits particles.

A maser emits electromagnetic radiation. This radiation has several measurable properties. (See diagram 1)
Frequency or wavelength is measured in hertz. Hertz is a
measurement of how many complete waveforms (as in
diag 1) occur in a one-second time frame. Amplitude
measures the voltage or power of the wave.

Masers emit radiation in the form of a concentrated
beam, just as a laser emits a concentrated beam of light.
The radiation leaving a maser does so at the speed of light
(186,282 miles per second!). The main difference between
a laser and a maser is the frequency at which they operate.
A maser operates in the microwave frequency range (about
300 billion hertz) and a laser operates in the frequency
range of light (About 300 trillion hertz).

Microwave's potential as a military weapon has been
kept secret from the general public since its creation.

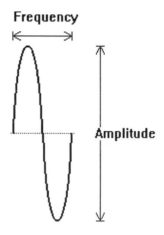

Lasers were talked about more because the individual components needed to make a laser of significant power were hard to come by. Magnetrons, however, were put into microwave ovens and sold to the public almost immediately. A commercial microwave oven has a power rating of 5000 watts. That is more than enough power to be classed "military."

When the electromagnetic radiation from a maser hits an electronic circuit, electric surges are created that can damage the components. The electromagnetic radiation, however, cannot pass through metal. When the rays hit metal they are reflected away, just as a laser bounces off a mirror. There will be sparks when microwave energy hits metal. A maser beam will pass through almost any other substance.

When maser energy hits a water molecule the molecule oscillates or vibrates extremely fast. The vibrations of molecules cause friction which in turn causes heat. This is how a microwave oven heats food. Anything a maser beam hits that contains water will get very hot, very fast.

The human body contains about 67 percent water. A maser beam would atrophy and sear living heart muscle in seconds. Likewise brain tissue would be permanently impaired almost instantly. Any other organs would be similarly affected in a very short period of time. No tests have been conducted in these areas that I am aware of, except those stories you hear of people putting cats in a microwave oven.

Wood products all contain water as well. Even kiln dried lumber has a 10% moisture content. When wood is hit by a maser, it gets hot until it combusts. The wattage of the maser, the distance between the maser and the target, and the moisture content of the wood determine how much time is required to achieve combustion. I have ignited a 4 x 8 sheet of plywood, from 500 feet away using a 5000 watt maser for only 30 seconds. A wood wall will not stop the maser beam from penetrating further and damaging the occupants or contents.

A maser placed in a satellite and powered by a small nuclear reactor could do significant damage anywhere on earth. Similarly, a large magnetron on earth could impair or destroy a satellite in orbit.

The Poor Man's Ray Gun

Chapter 2

How the Microwave Oven Works

The main component of a microwave is the magnetron. The magnetron is the maser that creates the electromagnetic radiation that cooks the food. In order to achieve even heating your oven needs that concentrated maser beam to be broken up and bounced around the oven cavity. Your oven does this by bouncing the radiation off of several round surfaces and over fan-like diffractors before the radiation enters the oven cavity. The part of the oven that houses these round surfaces and the fan diffractors is called the waveguide.

The next component is the control panel. There are usually two basic controls: time and power. The time circuits control how long or the duration power is applied to the magnetron. Since you can't actually adjust the power of your magnetron, the power switch just pulses your magnetron on and off for a length of time dependent upon the timer setting. We will not be using the control panel on our maser since it would require you to stand next to the maser when you fire it.

The door interlocks on your microwave oven ensure that the magnetron can not fire with the door open. There are two door switches to create redundancy for safety. The interlocks also control the light inside the oven cavity. We will not be incorporating door interlocks into our maser.

The cooling fan blows air across the magnetron to keep it from overheating. The wind it creates also turns the beam diffracting fan in the waveguide. We will be using the cooling fan in our maser.

The only other components we will cover are the fuse and the thermal protection device. The fuse protects the components from electrical short circuits by interrupting the incoming electricity. If the fuse blows you will have to replace it. The thermal protection device is mounted directly on the magnetron. It interrupts the incoming electricity if the magnetron gets too hot. Once the thermal protection device cools off it will turn the power back on.

Chapter 3
Taking it Apart

Selecting the microwave oven you will convert into a maser needs to be seriously considered. Some large commercial ovens require 220 volts of electricity to operate. Do you have 220 volts available where you want to use your maser? The higher wattage rating of the oven, the greater the effective operating range of the maser. Of course your budget needs to be considered.

You may want to check the dumpster of an appliance repair shop. Usually the magnetron and power supply circuits are fine, but it is in the dumpster because the timer, controls, or door integrity have failed. I built my first maser after I had a fire in my microwave oven. (you have to remove those metal straps from Chinese food containers) The oven was destroyed, but the magnetron and power supply were not damaged.

A basic overview of what we are about to do will be helpful. We will be removing the microwave's magnetron and related circuitry, while eliminating the door safety interlocks, timer, and power controls. We will then mount the magnetron and related circuitry to a new frame. You can build your maser if you can't read a schematic diagram, but we have included one for those people who can. (diagram 2) When we build our maser we will be eliminating all of the components between the dashed lines.

The Poor Man's Ray Gun

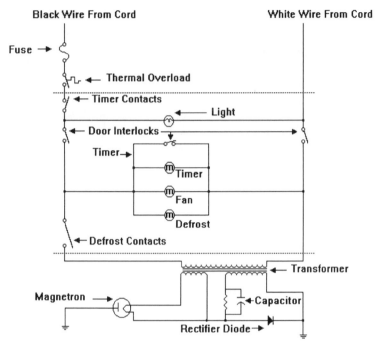

Black Wire From Cord
White Wire From Cord

Fuse →

Thermal Overload
Timer Contacts
Light
Door Interlocks
Timer→
Timer
Fan
Defrost
Defrost Contacts
Transformer
Magnetron →
Capacitor
Rectifier Diode→

Diagram - 2

Once you have obtained the oven to be dissected, be sure to disconnect the power, because very dangerous voltages are present with the magnetron's power supply. Even after the unit is unplugged high voltage may be present in the capacitor, so don't go sticking your hands inside until we identify and discharge the capacitor. Start by removing the outside cover from the microwave. The outside cover will not be reused so you can bend it as you remove it.

Once the cover is removed, carefully look to see if there is any documentation or schematic diagrams included with the oven. Most manufacturers include a whole packet of information hidden inside the oven cover. If they are included with your model they can help assist you in component identification.

The Poor Man's Ray Gun

We must now find and discharge the high voltage capacitor. (See photo 1) Again, do not go sticking your fingers or screwdrivers inside yet. Most capacitors will look very similar to this. Once you have it identified you should notice that the wires coming from it are large and have thick insulation around them. Some models will even have plastic shields around the terminals themselves to prevent arcing to other components.

Using well insulated pliers, pull the plastic shield up the wire away from the capacitor. Then take an insulated wire with one end connected to a good ground (such as a copper pipe) and take the other end of the wire and touch it to the capacitor terminals one at a time. Be careful and do not touch the pipe or the wire except for the insulated part. Also close your eyes just before you touch each terminal, in case there are sparks.

Most capacitors are already discharged before doing this procedure. It is good practice to do this each time you touch or go near the capacitor after the unit has been plugged in.

The next component to identify will be the magnetron. (see photo 2) Yours may not look exactly like this one, but it will be very similar. Also note the thermal protection device mounted on top of the magnetron. You will not be able to see the tip of the magnetron at this point as it is housed in the waveguide that carries the microwave energy to the oven cavity.

The cooling fan (see photo 3) should be the easiest component to identify. Notice that it is aimed directly at the magnetron.

We also need to find the transformer. (See photo 4) Most ovens have only one transformer. Some models do have a second transformer for the timer and clock in the oven. If the model you have has two transformers, we need to select the one with wires going directly to the magnetron and capacitor.

The power cord for your microwave oven will have three wires. The green one (ground) is connected directly

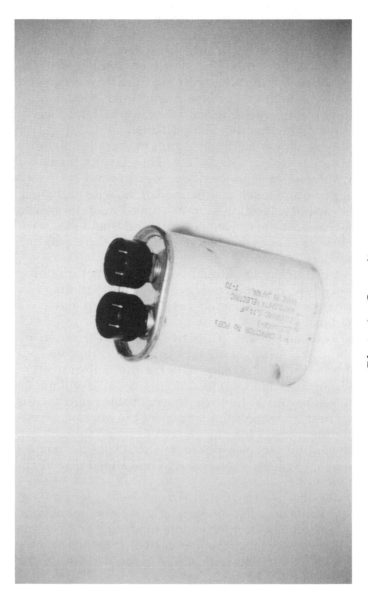

Photo 1 - Capacitor

The Poor Man's Ray Gun

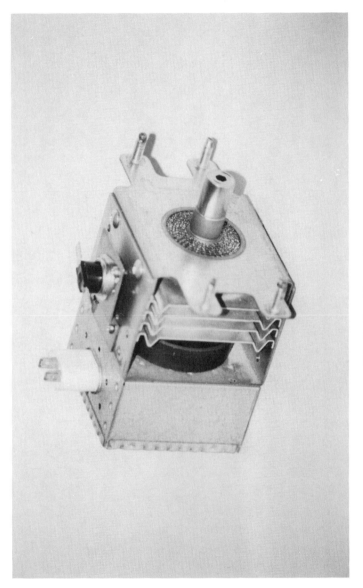

Photo 2 - Magnetron

Photo 3 - Cooling Fan

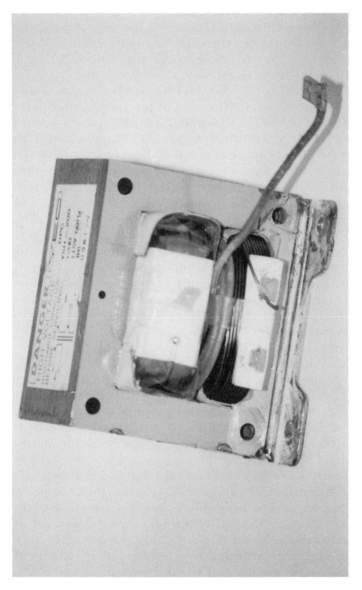

Photo 4 - Transformer

to the metal frame of the oven. The white wire (neutral) goes to the front control panel of the microwave. The black wire (hot) goes through a fuse assembly and then traditionally goes to the thermal protection device directly on the magnetron.

You should notice that your transformer has two sides with terminals on it. The side of the transformer that is connected to the control panel, and the power cord, is called the primary side, the side that is connected to the magnetron and capacitor is called the secondary side. The secondary side of your transformer may have two or three terminals, depending on the manufacturer. Most transformers do have a label indicating which terminals are primary and secondary.

Now you should draw a basic diagram of your main components, secondary side of the transformer, capacitor, magnetron, and any other components connected directly to any of the other main components, such as a rectifier diode (see photo 5) You may have to remove some wire ties to expose individual wires. Now draw the wires that connect the main components. Also describe any markings or coloring on the wires in your diagram.

You will have to disconnect the wires when you disassemble the oven so make every effort to mark the wires and terminals clearly so reassembly will be easy. I recommend putting a piece of masking tape near every terminal and on each end of each wire, so you can mark directly on every connection. You may for now disregard the thermal overload and fuse assembly.

Finally, once you are confident that you will be able to re-wire the main components, (transformer, capacitor, magnetron, and rectifier diode) begin disassembly. How your oven is configured will dictate what order you remove the parts. Remove the transformer, capacitor, rectifier diode, magnetron and any components wired directly to these components and rewire them back together as soon as each component is removed. Do not remove any components connected to the primary side of the transformer.

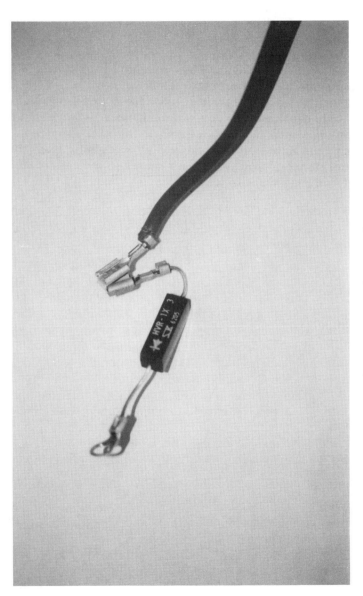

Photo 5 - Rectifier Diode

Be sure to save all of the mounting hardware you remove.
(Screws, nuts, etc.)

You should now remove the fan and disconnect its
wires from the control panel. Also remove the power cord,
fuse assembly and thermal overload.

You may incorporate the light from the microwave
oven into the maser, so you can see from a distance that it
is powered and on. If you wish to do this you will need to
remove the light bulb, fixture and wires coming from the
control panel. If you want your maser to look more dis-
creet you may want to leave it off.

Chapter 4
Putting it Together

The next step in constructing your maser is to design and build a frame. The frame can be very basic, (see photo 6) or the maser could be mounted in a briefcase. Where and how you will want to use the maser will determine most of the frame's configuration.

Now we will build a basic frame for the maser. Start with a piece of sheet metal about 1 foot x 2 foot. Arrange all of the main components (transformer, capacitor, magnetron) on the metal so that the magnetron's business end or nozzle is pointed away from the other components. Also make sure the existing wiring will reach each component because you don't want to splice high voltage wiring.

Your fan will have to be aimed directly at the magnetron's cooling fins. This will probably require that the fan be mounted to a vertical surface. You will need to make a 90 degree bend in the metal to create the vertical mounting surface. You can make this bend by placing a piece of 2 x 4 on top of your piece of metal. Then stand on the 2 x 4 and lift the metal up using the 2 x 4 as a guide for the fold. You will have to bend the metal a little past 90 degrees and it will spring back to 90 degrees.

Now with all of the parts laying in their proper position, (make sure the fan has clearance) mark each component's mounting holes with a magic marker or pencil.

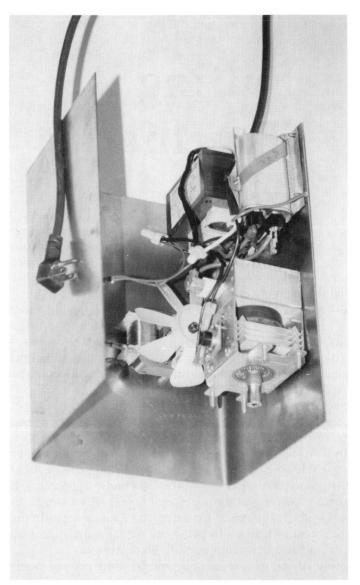

Photo 6 - Maser in a basic frame

The Poor Man's Ray Gun

Then, if you are mounting your fan vertically, lay the frame on its side and mark the fan mount holes.

Drill each hole you have marked with an appropriate sized drill. Make each hole a little bigger than the screw that will go through it. Then begin to mount the components to the frame, being careful not to stress any of the wires. Once you have everything mounted, go back over all the hardware and tighten it again. If you are going to use the light you should mount it now.

We will now begin to wire the maser. Do not plug in the unit to test it until you have read the rest of the book. This is a dangerous weapon and we have to cover maser safety.

The green wire from the power cord should be screwed directly to the metal frame of the maser. You can use a screw that holds down a component if you want.

The black wire should be connected to either end of the fuse holder. You should run a wire from the other end of the fuse holder to either terminal of the thermal overload. Now run a wire from the other end of the thermal overload to either terminal on the primary side of the transformer. Also, connect one of the two fan wires here as well as one of the two light wires.

The white wire should be connected to the open terminal on the primary side of the transformer. Also connect the other wire from the fan and the light to this terminal.

Now take some cardboard and tape, and construct a shield around the capacitor terminals, so that nothing can accidentally get near the terminals.

There, it is wired and almost ready to be tested. If you have any knowledge of electronics and electricity, you should have noticed that there is no on-off switch. If you plug it in, it will be on.

The Poor Man's Ray Gun

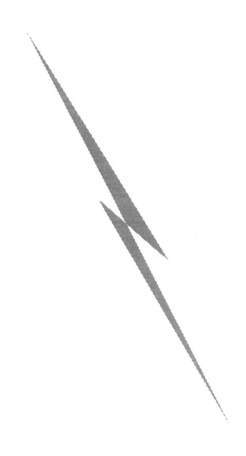

Chapter 5
Safety

Radio Shack sells an appliance remote control system, which enables you to switch the maser on and off at a great distance from the maser. Using a remote switching system is the safest way to fire your maser.

If you do not wish to use the remote system, you can use an extension cord plugged into a switched receptacle. Make sure you use an extension cord with a ground, because you may not bypass the ground.

Whatever you use to turn your maser on, make sure you are a considerable distance from the maser. Make sure that the energy is not aimed at metal that will reflect the radiation back to your location. Remember, the electro-magnetic radiation is traveling at the speed of light, so if it bounces around for only 2 seconds it has traveled extremely far. If possible have a metal wall between you and the maser.

Another good safety idea is to aim a video camera at the maser so you can view the firing remotely. Most video cameras can be hooked up directly to a television. You will need a good length of cabling. Just be sure the camera is not going to be hit by the maser directly.

To test your maser, put a small cup of water in front of the magnetron, about 4 inches away. Then fire the maser for about 5 seconds. You will notice that the water is much hotter than it would have been had it been placed in the microwave oven for the same time. This is because the microwave oven breaks up the maser beam so the oven heats more evenly.However, here we are dealing with the concentrated pure beam.

Lastly, be sure that whenever you power up your maser, nobody is downrange that you don't intend to be downrange. Remember, you can kill from a great distance with a maser.

Chapter 6
Strategic Uses

Masers can be used as a defensive or offensive weapons system. Using several masers together in an array could be used to protect or invade any given area.

If you had an entryway you needed to protect, a maser can be a good choice. If there was sufficient overhead you could enclose the maser in an attic aimed through a small hole straight down. Again, I have not conducted any tests on any living matter. But I can't imagine a brain being very useful after receiving a 10 second exposure from a 2500 watt magnetron at two feet away. You could hook the maser up to a 120 volt output on an alarm system. Then, rather than a siren going off and scaring the burglar, you have criminal flambé. Please check your local laws regarding active alarm systems. Also be sure your system doesn't false alarm and cook a relative or friend.

You could use a good DC to AC converter that plugs into your car's cigarette lighter. This can be used to power your maser while away from home. Just be certain the beam clearly exits the car. If it starts bouncing around inside the vehicle the results would be disastrous (Unless your target happens to be in the vehicle).

It would be possible to disable a vehicle, if you can get the microwave energy to reflect through the radiator into the engine compartment. The energy will not pass through the metal quarterpanels. A maser mounted under a bridge or under a road would disable a vehicle as it passed over. If the vehicle has enough momentum it may restart after it passed the maser. If the car has an electronic ignition the components will probably be fried.

Whatever plant material your maser hits will die immediately. It will take a week for the leaves to turn brown and fall off, but it will die immediately. This is important to remember if you plan to shoot your maser at someone's house, you will mark your point of entry in the bushes. Marking a tree or bush can assist you in long distance aiming.

Using a maser to protect a long tunnel leading to an underground bunker would be a very effective defense. The tunnel would fill up with bodies before anyone could get you out of the bunker.

Notes

Other Books Available From Desert Publications

001	Firearms Silencers Volume 1	$9.95
003	The Silencer Cookbook	$9.95
004	Select Fire Uzi Modification Manual	$9.95
005	Expedient Hand Grenades	$16.95
007	007 Travel Kit, The	$8.00
008	Law Enforcement Guide to Firearms Silencer	$8.95
009	Springfield Rifle, The	$11.95
010	Full Auto Vol 3 MAC-10 Mod Manual	$9.95
012	Fighting Garand, The	$11.95
013	M1 Carbine Owners Manual	$9.95
014	Ruger Carbine Cookbook	$11.95
015	M-14 Rifle, The	$9.95
016	AR-15, M16 and M16A1 5.56mm Rifles	$11.95
017	Shotguns	$11.95
019	AR15 A2/M16A2 Assault Rifle	$8.95
022	Full Auto Vol 7 Bingham AK-22	$9.95
027	Full Auto Vol 4 Thompson SMG	$9.95
030	STANAG Mil-Talk	$9.95
031	Thompson Submachine Guns	$13.95
033	H&R Reising Submachine Gun Manual	$12.95
035	How to Build Silencers	$6.95
036	Full Auto Vol 2 Uzi Mod Manual	$9.95
049	Firearm Silencers Vol 3	$19.95
050	Firearm Silencers Vol 2	$19.95
054	Company Officers HB of Ger. Army	$11.95
056	German Infantry Weapons Vol 1	$14.95
058	Survival Armory	$27.95
060	Survival Gunsmithing	$9.95
061	FullAuto Vol 1 Ar-15 Mod Manual	$8.95
064	HK Assault Rifle Systems	$27.95
065	SKS Type of Carbines, The	$16.95
066	Private Weaponeer, The	$9.95
067	Rough Riders, The	$24.95
068	Lasers & Night Vision Devices	$29.95
069	Ruger P-85 Family of Handguns	$14.95
071	Dirty Fighting	$11.95
072	Live to Spend It	$29.95
073	Military Ground Rappelling Techniques	$11.95
074	Smith & Wesson Autos	$27.95
080	German MG-34 Machinegun Manual	$9.95
081	Crossbows! From 35 Years With the Weapon	$11.95
082	Op. Man. 7.62mm M24 Sniper Weapon	$7.95
083	USMC AR-15/M-16 A2 Manual	$16.95
084	Urban Combat	$7.95
085	Caching Techniques of U.S. Army Special Forces	$9.95
086	US Marine Corps Essential Subjects	$16.95
087	The L'il M-1, The .30 Cal. M-1 Carbine	$14.95
088	Concealed Carry Made Easy	$14.95
089	Apocalypse Tomorrow	$7.95
090	M14 and M14A1Rifles and Rifle Marksmanship	$11.95
091	Crossbow As a Modern Weapon	$11.95
092	MP40 Machinegun	$11.95
093	Map Reading and Land Navigation	$9.95
094	U.S. Marine Corps Scout/Sniper Training Manual	$14.95
095	Clear Your Record & Own a Gun	$14.95
096	Sig Handguns	$14.95
097	Poor Man's Nuclear Bomb	$16.95
098	Poor Man's Sniper Rifle	$19.95
100	Submachine Gun Designers Handbook	$16.95
101	Lock Picking Simplified	$8.50
102	Combination Lock Principles	$8.95
103	How to Fit Keys by Impressioning	$9.95
104	Keys To Understanding Tubular Locks	$9.95
105	Techniques of Safe & Vault Manipulation	$9.95
106	Lockout - Techniques of Forced Entr	$11.95
107	Bugs Electronic Surveillance	$11.95
110	Improvised Weapons of Amer. Undergrnd	$10.00
111	Training Handbook of the American Underground	$10.00
114	FullAuto Vol 8 M14A1 & Mini 14	$9.95
116	Handbook Bomb Threat/Search Procedures	$8.00
117	Improvised Lock Picks	$10.00
119	Fitting Keys By Reading Locks	$7.00
120	How to Open Handcuffs Without Keys	$9.95
121	Electronic Locks Volume 1	$8.00
122	With British Snipers, To the Reich	$24.95
125	Browning Hi-Power Pistols	$9.95
126	P-08 Parabellum Luger Auto Pistol	$9.95
127	Walther P-38 Pistol Manual	$9.95
128	Colt .45 Auto Pistol	$9.95
129	Beretta - 9MM M9	$11.95
130	FullAuto Vol 5 M1 Carbine to M2	$9.95
132	Ranger Handbook	$16.95
133	FN-FAL Auto Rifles	$13.95
135	AK-47 Assault Rifle	$11.95
136	Uzi Submachine Gun	$11.95
139	USMC Battle Skills Training Manual	$24.95
140	Sten Submachine Gun, The	$9.95
141	Terrorist Explosives Handbook	$6.95
142	U.S. Army Counterterrorism Training Manual	$24.95
143	Sniper Training	$24.95
144	The Butane Lighter Handgrenade	$9.95
146	The Official Makarov Pistol Manual	$12.95
152	Glock's Handguns	$17.95
153	Heckler and Koch's Handguns	$17.95
154	The Poor Man's Ray Gun	$9.95
200	Fighting Back on the Job	$11.95
202	Secret Codes & Ciphers	$7.95
204	Improvised Munitions Black Book Vol 1	$14.95
205	Improvised Munitions Black Book Vol 2	$14.95
206	CIA Field Exp Preparation of Black Powder	$8.95
207	CIA Field Exp. Meth/Explo. Preparat	$8.95
209	CIA Improvised Sabotage Devices	$12.00

210	CIA Field Exp. Incendiary Manual	$12.00
211	Science of Revolutionary Warfare	$9.95
212	Agents HB of Black Bag Ops	$9.95
214	Electronic Harassment	$11.95
217	Improvised Rocket Motors	$6.95
218	Impro. Munitions/Ammonium Nitrate	$9.95
219	Improvised Batteries/Det. Devices	$8.95
220	Impro. Explo/Use In Deton. Devices	$7.95
222	Evaluation of Imp Shaped Charges	$8.95
224	American Tools of Intrigue	$12.00
225	Impro. Munitions Black Book Vol 3	$25.95
226	Poor Man's James Bond Vol 2	$24.95
227	Explosives and Propellants	$11.95
229	Select Fire 10/22	$11.95
230	Poor Man's James Bond Vol 1	$24.95
231	Assorted Nasties	$19.95
234	L.A.W. Rocket System	$8.00
240	Clandestine Ops Man/Central America	$11.95
241	Mercenary Operations Manual	$9.95
250	Improvised Shaped Charges	$8.95
251	Two Component High Exp. Mixtures	$9.95
260	Survival Evasion & Escape	$13.95
262	Infantry Scouting, Patrol, & Sniping	$13.95
263	Engineer Explosives of WWI	$7.95
300	Brown's Alcohol Motor Fuel Cookbook	$13.95
301	How to Build a Junkyard Still	$11.95
303	Alcohol Distillers Handbook	$17.95
306	Brown's Book of Carburetors	$11.95
310	MAC-10 Cookbook	$9.95
350	Cheating At Cards	$12.00
367	Brown's Lawsuit Cookbook	$18.95
400	Hand to Hand Combat	$11.95
401	USMC Hand to Hand Combat	$7.95
402	US Marine Bayonet Training	$8.95
408	Camouflage	$12.95
409	Guide to Germ Warfare	$13.95
410	Emergency War Surgery	$24.95
411	Homeopathic First Aid	$9.95
412	Defensive Shotgun	$12.95
414	Hand to Hand Combat by D'Eliscue	$7.95
415	999 Survived	$6.00
416	Sun, Sand & Survival	$6.00
420	USMC Sniping	$16.95
424	Prisons Bloody Iron	$13.95
425	Napoleon's Maxims of War	$9.95
429	Invisible Weapons/Modern Ninja	$11.95
432	Cold Weather Survival	$11.95
435	Homestead Carpentry	$10.00
436	Construction Secret Hiding Places	$11.95
437	US Army Survival	$29.95
438	Survival Shooting for Women	$9.95
440	Survival Medicine	$11.95
442	Can You Survive	$12.95
444	Canteen Cup Cookery	$9.95
445	Leadership Hanbook of Small Unit Ops	$11.95
447	Vigilante Handbook	$9.95
453	Shootout II	$14.95
454	Catalog of Military Suppliers	$14.95
455	Survival Childbirth	$8.95
456	Police Karate	$10.95
457	Survival Guns	$21.95
470	Water Survival Training	$5.95
475	Emergency Medical Care/Disaster	$11.95
500	Guerilla Warfare	$9.95
504	Ranger Training & Operations	$14.95
507	Spec. Forces Demolitions Trng HB	$16.95
508	IRA Handbook	$9.95
510	Battlefield Analysis/Inf. Weapons	$9.95
511	US Army Bayonet Training	$9.95
512	Desert Storm Weap. Recog. Guide	$9.95
542	Professional Homemade Cher Bomb	$10.95
551	Combat Loads for Sniper Rifles	$12.95
552	Take My Gun, If You Dare	$10.95
605	Aunt Bessie's Wood Stove Cookbook	$7.50
610	Jackedup & Ripped off	$10.00
C-002	Trapping & Destruc. of Exec. Cars	$10.00
C-011	How To Open A Swiss Bank Account	$6.95
C-019	Defending Your Retreat	$9.95
C-020	How To Become A Class 3 MG Dealer	$35.95
C-021	Methods of Long Term Storage	$8.95
C-022	How To Obtain Gun Dealer Licenses	$8.95
C-023	Confid. Gun Dealers Guide/Whlesler	$8.95
C-028	Federal Firearms Laws	$4.50
C-030	Militarizing the Mini-14	$7.95
C-038	M1 Carbine Arsenal History	$6.95
C-040	OSS/CIA Assassination Device	$4.00
C-050	Hw To Build A Beer Can Morter	$4.00
C-050	Criminal Use of False ID	$11.95
C-058	Surviving Doomsday	$11.95
C-080	CIA Explosives for Sabotage	$9.00
C-085	Brass Knuckle Bible	$9.00
C-099	USA Urban Survival Arsenal	$7.95
C-175	Dead or Alive	$7.95
C-177	Elementary Field Interrogation	$7.95
C-209	Beat the Box	$7.95
C-386	Boobytraps	$8.00
C-979	Guide/Vietcong Boobytraps/Device	$8.00
C-9031	Self-Defense Requires No Apology	$11.95
FP-9	M16A1 Rifle Manual Cartoon Version	$6.95
	Become a Licensed Gun Collector	$4.95
	Micro Uzi Select Fire Mod Manual	$9.95

PRICES SUBJECT TO CHANGE WITHOUT NOTICE

Send $5.00 for a complete 72 page catalog Free with order

DESERT Publications
P.O. Box 1751 Dept. BK-155
El Dorado, AR 71730-1751 USA

Add $5.95 to all orders for shipping & handling.